STUDENT GUIDE

MATHEMATICS of MOTION

DISTANCE, SPEED, AND TIME

MathScape™
SEEING AND THINKING MATHEMATICALLY

How can mathematics be used to analyze motion?

MATHEMATICS *of* MOTION

PHASE**ONE**

Measuring, Estimating, and Representing Motion

In this phase, you will practice measuring motion and estimating how fast you move, how long it takes, and how far you go. Making good measurements is an important part of this beginning study of motion. You will also find new ways to represent motion in words, pictures, and diagrams.

PHASE**TWO**

Distance-Time Graphs

First, you will walk at different speeds to collect motion data. Then you will use your data to make graphs that show how far you moved versus how long it took—distance-time graphs. You will also compare graphs with maps and stories that describe motion to find the best way to represent motion in any situation.

PHASE**THREE**

Using Graphs to Solve Problems

For more practice graphing, you will convert a sports announcer's race commentary into a graph. You will learn about average speed and find out it might be different than speed at any particular time. You will also find out about a powerful mathematical expression for calculating distance, speed, or time of travel. A final project brings together all of your new skills in dealing with motion.

PHASE ONE

To: Future Olympic Runners

To improve your performance, we on the coaching staff want you to learn more about motion. The more you know, the more you can track your progress. The first thing you need to do is to take accurate measurements of how far you run. You also need to find out how you can determine speed if you know your distance and time. Because you may not always be able to measure your motion, you should also be able to make accurate estimates of distance, time, and speed. We will want to discuss your training sessions, so it's important to know how to describe and represent your motion too.

Get ready to hop, skip, and jump! In this phase you will learn how the motions that we make can be analyzed and described with mathematics. Studying motion is not only fun, it can be an important part of your job if you are an engineer, astronaut, or athlete, for example.

Measuring, Estimating, and Representing Motion

WHAT'S THE MATH?

Investigations in this section focus on:

MEASUREMENT and GEOMETRY

- Estimating and measuring distance, time, and speed
- Using units of measurement

NUMBER and COMPUTATION

- Plotting positions of parade characters on a Snapshot Sequence Sheet
- Using a Snapshot Sequence Sheet to determine direction, change in speed, and distance traveled over an interval

ALGEBRA FUNCTIONS

- Determining distance, time, or speed when the other two measures are known

MathScape Online
mathscape3.com/self_check_quiz

1 Moving, Measuring, and Representing

INVESTIGATING RATES OF MOTION

Fast or slow, up or down, turning, twisting, or straight—there are many kinds of motion. You will move your body in different ways, estimate and measure speed and distance, and use different methods to represent motion.

Move at Precise Speeds

How closely can you estimate the speed of motion?

In the following activities, work with your partner. Take turns being the mover and the measurer.

1. The mover tries to move at exactly the speed given in the List of Motions without looking at a clock or watch. The measurer measures, times, and checks the mover.
2. Make up new motions at precise speeds. Choose what the mover should do and how fast the mover should go. Record what motions you tried and what happened.
3. Write a description of something that might move at each of the following speeds:

   a. 5 feet per second
   b. 3 meters per second
   c. 50 meters per hour
   d. 10 miles per hour
   e. 50 miles per hour
   f. 1,200 miles per hour

List of Motions

Tap a pencil at exactly 1 tap per second for 10 seconds.

Turn the pages of a book at exactly 2 pages per second for 10 seconds.

Move at a speed of exactly 1 foot per second for 30 seconds.

Walk from your seat to the classroom door at a speed of 2 feet per second.

Tap a pencil every 5 seconds for 1 minute.

Turn the pages of a book at exactly 1 page per second for 20 seconds.

Move at a speed of exactly 1 meter per second for 10 seconds.

Walk the length of the classroom in exactly 15 seconds. What was your speed?

Represent Your Trip to School

How can words and visual images be used to represent motion?

If you need to explain how something moves, there are different kinds of representations to choose from. For example, consider how you can represent the motion of your daily trip to school.

1 Describe how you get to school in writing. You can use a story, a set of directions, or any other way to describe your trip in words. Make sure someone could tell from your description whether you walk or ride to school, how long it takes you to get there, and how far and how fast you travel.

2 Show your trip to school visually. You can use a map, a diagram, a graph, pictures, or any other kind of visual representation.

3 Use the list Things to Know About Your Trip to answer the following two questions.

a. What could someone learn about your trip by reading your description of it?

b. What could they learn by looking at your visual representation?

Things to Know About Your Trip

- Each turn that you make along the way
- The names of the streets
- The mode of transportation (walk, bicycle, bus, car, train)
- The distance of the entire trip
- The distance of each part of the trip
- The time of the entire trip
- The time of each part of the trip
- Whether you go with anyone else
- The speed you travel for each part of the trip

hot words | estimate

Homework
page 78

2 Going the Distance

ESTIMATING, MEASURING, AND CALCULATING DISTANCE

To find out how far it is from one point to another, do you estimate the distance? measure it? calculate it? In this lesson, you will have the chance to use all three ways of finding distance.

When you move from here to there, how do you know how far you have gone?

Find Out How Far

1 Copy the table Knowing the Distance in your notebook or on another sheet of paper. Without actually walking or measuring, estimate how far you will go in 10 seconds when you move each way listed in the table. Record your estimates in your table.

2 Working with your partners, measure how far you actually go in 10 seconds for each way of moving. How do your estimates compare with your measurement results?

3 Using the data from the second column, calculate how far you would go in 1 second, 1 minute, and 1 hour; and how long it would take you to go 1 mile or 1 kilometer.

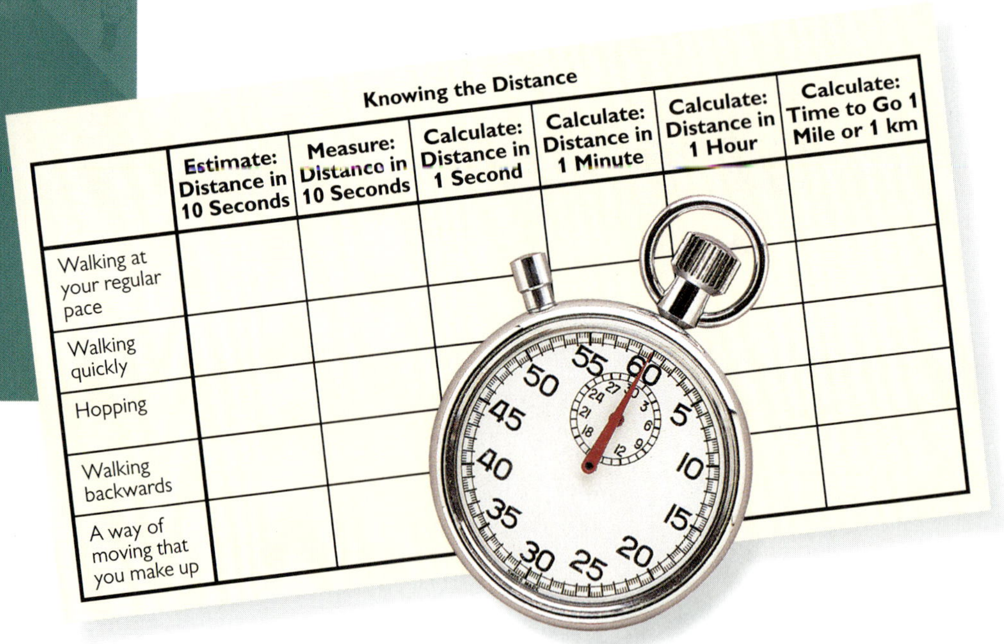

Knowing the Distance

	Estimate: Distance in 10 Seconds	Measure: Distance in 10 Seconds	Calculate: Distance in 1 Second	Calculate: Distance in 1 Minute	Calculate: Distance in 1 Hour	Calculate: Time to Go 1 Mile or 1 km
Walking at your regular pace						
Walking quickly						
Hopping						
Walking backwards						
A way of moving that you make up						

Relate How Long? How Far? and How Fast?

Copy the table Going Places in your notebook or on a sheet of paper. Include all the information already given. Then fill in the rest of your copy of the table. Each row should show places you could start from and go to, how you could get there (by foot, by bike, by bus, etc.), and the distance, time, and speed of the trip. You may have to estimate some of the information. Some of it you will find by making calculations.

Is there a relationship between time, distance, and speed?

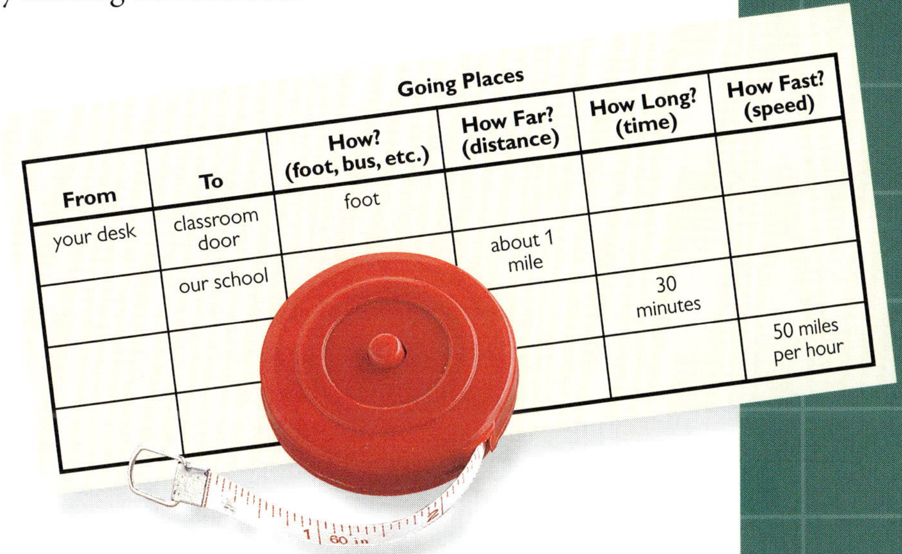

Going Places

From	To	How? (foot, bus, etc.)	How Far? (distance)	How Long? (time)	How Fast? (speed)
your desk	classroom door	foot			
	our school		about 1 mile		
				30 minutes	
					50 miles per hour

Explain Your Solutions

After you have completed the table, choose one distance, one time, and one speed that you were not able to measure directly. Describe in writing why you couldn't measure it directly and how you figured out the numbers you put in the table.

hot words
equation
distance

Homework
page 79

3 Reporting Live from the Parade

REPRESENTING MOTION WITH PICTURE COORDINATES

Have you ever seen the film used in movie projectors? Each frame, or "snapshot," shows the scene at a different time. Taken together, the snapshots show action. You can use written snapshots to create a visual representation of the action at a parade.

How can snapshots be used to show the motions of parade characters?

Use Snapshots to Represent Motion

When you read the handout Reporting Live from the Parade, you learned that Monica Chang is a radio reporter covering a parade. JT Diaz, the photographer, takes a snapshot every 5 seconds, capturing the changing positions of parade characters. Latasha Williams is JT's assistant. She takes detailed notes describing the scenes in JT's photos.

1. Use the information in the handout Latasha's Field Notes I, snapshots 1–8, to complete a Snapshot Sequence Sheet. For each snapshot, fill in the snapshot number and the time.

2. Record the position of the rollerblader, the drum major, and the dragon float when JT took snapshots 1–8. Choose a letter, symbol, or color to represent each parade character.

Practice Problem Solving with the Snapshots

Read the following bulletin just in from the parade:

Live from the Parade

"This is Monica Chang again, from KMTH radio, bringing you more live coverage from the Main Street parade. JT has been taking pictures during our commercial break, and he tells me he's getting ready to take his 17th picture. The dragon float is still in view. We can also see a juggler on a unicycle and Lisa, a photographer friend of JT's from the newspaper. Lisa is jogging with her camera. That is unusual. I wonder why she's in such a hurry?"

- Make another visual representation of the action at the parade. This time, use the handout Latasha's Field Notes I for snapshots 17–24 and another Snapshot Sequence Sheet.
- Answer the question, "Why is Lisa in such a hurry?"

Write Equations for Motion

Use your Snapshot Sequence Sheets to write equations describing motion at the parade.

1 Write a rule you can use to figure out how many snapshots JT would take in any given number of minutes. Use P to stand for the number of snapshots, and M to stand for the number of minutes.

2 Assuming the drum major moves at a constant rate, write a rule you can use to figure out how far in meters the drum major would march in a given number of minutes. Use D to stand for the distance in meters he marches and M to stand for the number of minutes.

3 Assuming they move at constant rates, write rules to find the distance in meters the rollerblader and the dragon float move in any given number of minutes.

hot words
coordinates
equation

HW Homework
page 80

4 The Parade Continues

FINDING DATA FROM A SNAPSHOT SEQUENCE

Things have changed at the parade. There are new characters on the scene, and Monica Chang, the "Eye in the Sky," is having a hard time making her news report. Can you help Monica figure out what's going on?

Find Missing Data

How can you determine unknown positions on a snapshot sequence?

Read Monica's report below, Reporting Live from the Clouds. Then read the handout Latasha's Field Notes II. Notice that clouds prevented Latasha from taking notes some of the time. On a Snapshot Sequence Sheet, fill in the picture number and the time for each photograph. Then record the positions of the tumbling team, the clown, and the pig in each snapshot. What would JT's pictures have looked like if the clouds had not blocked the view in the 4th, 5th, 6th, and 8th pictures?

Reporting Live from the Clouds

"This is Monica Chang, again, bringing you live coverage of the parade from our Eye on the Scene hot air balloon. We're still looking down on Main Street, between Avenue A and Avenue B. JT has just loaded his camera with a new roll of film, and we're awaiting the next group of parade characters to come into our view."

"Now, JT, it's gotten pretty cloudy all of a sudden," says Monica, "Do you think you'll be able to take any good pictures?"

"I hope so, Monica. The clouds are going to make it difficult for me to take pictures, and for Latasha to see what's going on well enough to take notes, but they seem to be moving in and out, so I should be able to get a clear view for at least some of the pictures."

"Well, JT, get your camera ready. I can just barely see a clown on a skateboard approaching the Gas Station, and just behind her I think I see the Twisted Toes Tumbling Team turning cartwheels. They must be getting dizzy! And...what's that? There's a pig heading from the Stop Sign at Avenue B toward the Gas Station! That's strange. Maybe it escaped from the Creature Corral Petting Farm float."

Imagine Using a Faster Camera

Assuming the characters are moving at constant rates, what happened to the pig and the clown between 1:04:15 P.M. and 1:04:20 P.M.? Imagine that you have a faster camera than JT, and you can take a picture every 1 second between 1:04:15 and 1:04:20.

1. Create a snapshot sequence with a picture every 1 second.
2. Write an explanation of what happened to the pig and the clown between 1:04:15 P.M. and 1:04:20 P.M.

What happens between the snapshots in a snapshot sequence?

Write Your Own Snapshot Story

Write a story explaining the snapshot sequence that your teacher gives you. Use your imagination. Your story can be about anything you want!

1. Before you write your story, do the following:
   - **a.** Decide what the story is about. Who or what are Characters A, B, and C? What person, place, or thing is at each marker along the road?
   - **b.** Decide how much time there is between each snapshot and fill in the time of day that each snapshot was taken.
   - **c.** Decide how much distance there is between each of the 5 landmarks on the road or path.
2. Your story should also include information about the directions and speeds of the characters, and why they are moving that way.

hot words | distance

Homework
page 81

PHASE TWO

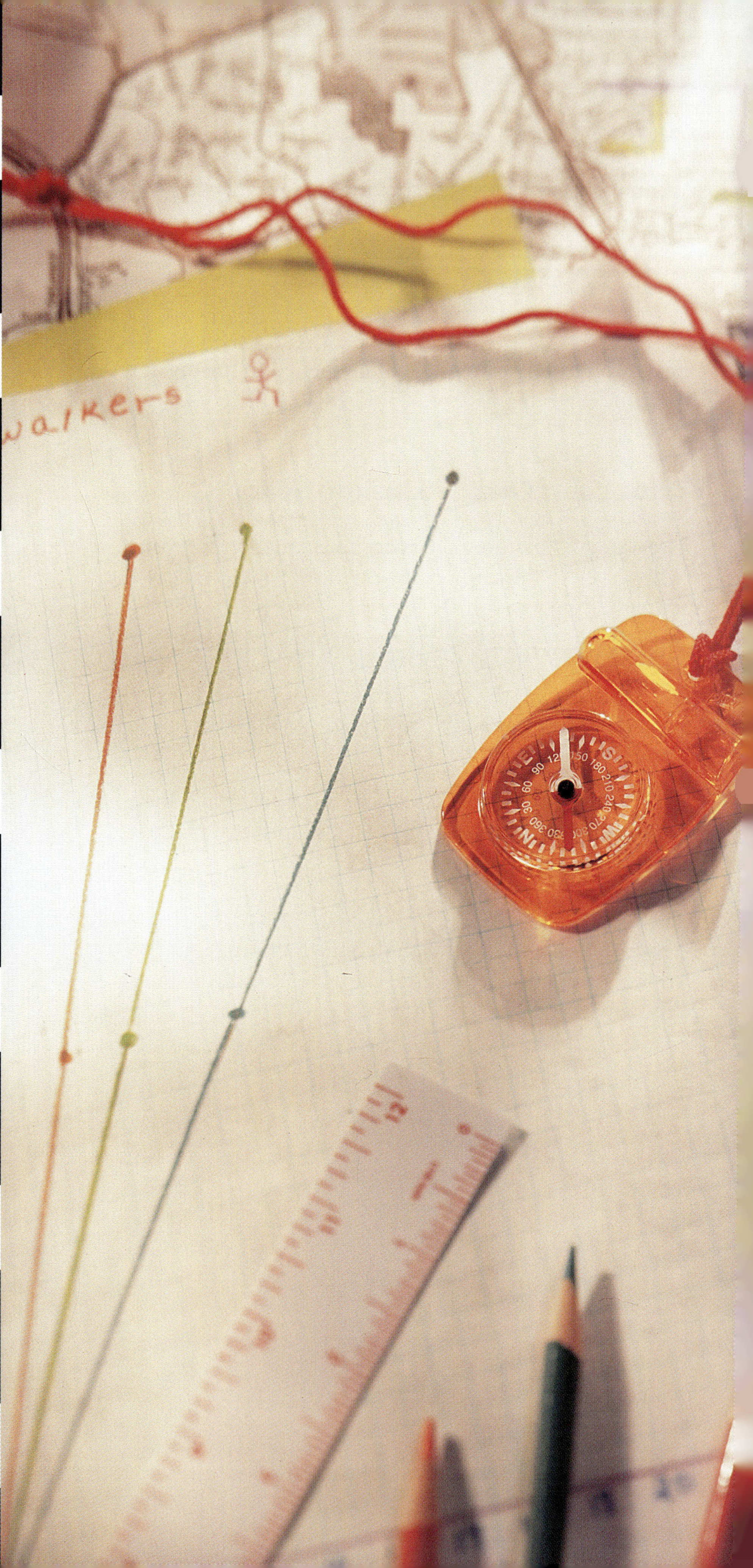

Distance-time graphs are powerful: they give you a lot of information in a very compact form. With a distance-time graph, you can quickly see how long it took someone to go someplace, how fast they went, and how far they traveled. You can see, at a glance, which contestant won a race, who lost, and who didn't finish. You can see whether everyone started at the same time, or if somebody had a head start. The distance-time graph is one mathematical tool that belongs in everyone's mathematical tool kit.

Distance-Time Graphs

WHAT'S THE MATH?

Investigations in this section focus on:

MEASUREMENT and GEOMETRY

- Setting appropriate scales on graphs
- Interpreting slopes on graphs

NUMBER and COMPUTATION

- Creating and interpreting distance-time graphs
- Relating different types of distance-time graphs to each other and to stories
- Reading information from maps

ALGEBRA FUNCTIONS

- Finding distance, time, and speed data in stories

MathScape Online
mathscape3.com/self_check_quiz

5 Walk This Way

INTRODUCING DISTANCE-TIME GRAPHS

As you have discovered, motion can be represented visually in a variety of ways. There are drawings and maps, and diagrams such as the snapshot sequences you created in Lessons 3 and 4. A graph is another way to represent motion visually.

Create Distance-Time Graphs

How can you make graphs representing walks at different speeds?

Apply what you learned about making distance-time graphs by graphing your data from the three walks at different speeds.

1. Collect data to graph by working with partners to measure the speeds of a slow walk, a regular walk, and a fast walk.
   - **a.** Decide who will walk and who will measure each walk. You must also measure the distance walked and the time it takes in order to calculate the speed.
   - **b.** Copy the table, Walks of Various Speeds, to store your data.
2. Make three distance-time graphs—one for each of the slow, regular, and fast walks. Label the axes to show the distance and time measures.

Walks of Various Speeds

	Distance	Time	Speed
Slow			
Regular			
Fast			

Connect Stories and Graphs

How can you make a distance-time graph to represent the activities of characters in a story?

Sometimes, data is gathered from an experiment before it is graphed. Other times, graphs can be prepared from information given in a story.

1. Choose any three of the characters in the parade in Lessons 3 and 4 and make a distance-time graph showing how they moved. Remember to label the axes to show the distance and time measures.

2. Read the story "Tanya and Pedro's Trips to School." Create a graph to show what happens in the story. Be sure your graph shows about how far each person went and how much time they spent.

Tanya and Pedro's Trips to School

Tanya's house is 1 mile away from her school. She started out to school at 8:00 A.M. On her way to school, Tanya walked for 5 minutes, going about $\frac{1}{4}$ of a mile. She then spent 5 minutes waiting for her friend Pedro. Pedro wasn't ready, and Tanya was worried that she would be late for school. She walked for 15 more minutes to get to school.

Pedro started from his house and ran to school in 10 minutes. He got there at the same time as Tanya arrived.

Relate Graphs to Motion

Write your answers to these questions about the distance-time graphs you made.

- How did you decide what scale to use on the axis of each of your graphs?
- If you used a different scale, how would your graph change? How would it stay the same?
- Are all the lines on your graphs straight? Why or why not?
- When two lines are on the same graph, what does it tell you when one is steeper than the other? Why?

hot words
coordinate graph
point

Homework page 82

6 Stories, Maps, and Graphs

COMPARING AND CONTRASTING REPRESENTATIONS

As you know, there are several useful ways to represent information about motion. Stories, maps, and graphs are some of these ways. But do stories, maps, and graphs provide the same information? Are they all equally useful?

Interpret a Story, Map, and Graph

Read the story, map, and graph about Jessica's travels. Then answer the questions on the sheet your teacher will give you that asks what Jessica was doing.

Information About Jessica's Travels

Jessica lives at the corner of 1st Street and Avenue A. She was planning to meet her friends at 10:30 A.M. at the school.

At 10:00 A.M., her mother asked her to go to the corner grocery store to get milk for her brother. She jumped on her bike and rushed out, biking two blocks along Avenue A and then going on 3rd Street to the grocery store.

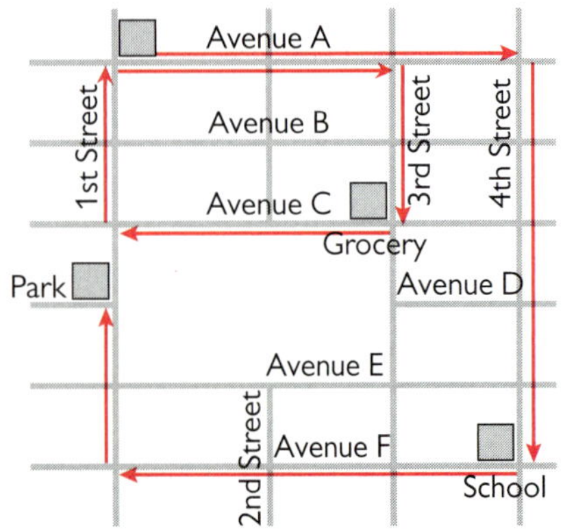

A Map of Jessica's Travels

She grabbed the milk and went to the express line. There were several people in front of her, so she had to wait.

When she left the grocery store, she discovered she had a flat tire. She walked her bike home along Avenue C and 1st Street. When she got home, she had to wait for her brother to drink his milk. Then her mother drove her to school, but she was late, and her friends weren't there. Her mother drove her to the park to look for her friends. She finally caught up to her friends at the park at 11:00 A.M.

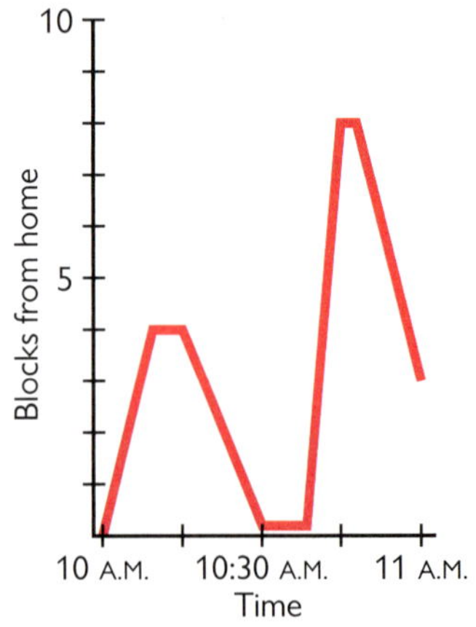

A Graph of Jessica's Travels

Graph Other Things That Change Over Time

Graphs can be used to show how an object's position—or almost anything else—changes over time. Draw a graph that shows how Jessica felt during different times in the story. You can graph how happy she felt, how annoyed, or any feeling you choose.

What kinds of changes over time can be shown on a graph?

Create Stories, Maps, and Graphs

Choose a trip that you are familiar with, or make up a fantasy trip. For your trip, do each of the following.

- Write a description or story.
- Draw a map.
- Make a distance-time graph.
- Make one other graph showing how you or a character in your story felt during the trip.

What kinds of travels can you describe with a story, map, and graph?

hot words | graph
picture graph

Homework
page 83

7 A Graphing Matter

MAKING THREE TYPES OF DISTANCE-TIME GRAPHS

Distance-time graphs come in several different sorts. You've had some experience with graphs that show how far something is from its starting point. Now you will learn how to make two other useful distance-time graphs.

What are some useful kinds of distance-time graphs?

Make Three Types of Distance-Time Graphs

In this activity, you and your partners will make three different kinds of graphs—Graph A, Graph B, and Graph C. Either your teacher or someone in your group will be the mover.

1 The mover does the following: Select a place where you can walk in a straight line while the others watch. When walking, walk at a steady pace. Others will need to count the number of steps you take, so don't go too quickly. When everyone is ready:

- **a.** Walk from the starting point to the ending point.
- **b.** Wait for about 5 seconds.
- **c.** Walk halfway back to the starting point.
- **d.** Wait about 5 seconds.
- **e.** Walk back to the ending point.
- **f.** Wait about 5 seconds.
- **g.** Walk back to the starting point.

2 Make the graphs. Each member of the group should make a different graph—either A, B, or C. Each graph will show something different about the number of steps the mover has taken. Graph A shows how many steps the mover is from the starting point throughout the walking time. Graph B shows how many steps the mover is from the ending point throughout the walking time. Graph C shows the total number of steps the mover has taken throughout the walking time.

Graph New Routes

Practice making the three kinds of distance-time graphs with new walking paths.

1 Decide who will be the mover and what the mover's path will be. If your teacher doesn't provide a path, make one up. Here is an example:

- **a.** Start at the door to the classroom.
- **b.** Walk slowly to the teacher's desk.
- **c.** Wait 10 seconds.
- **d.** Walk quickly to the bookcase.

2 Decide which graph you will make. Make either an A graph (distance from the starting point), a B graph (distance to the end point), or a C graph (total distance).

3 Watch the walker walk the path and gather the data you need to make your graph. Make your A, B, or C graph.

4 Repeat the process twice more but rotate which graph you make, so that each member of your group makes each kind of graph.

Can you make three kinds of distance-time graphs for new walking routes?

Convert Among Types of Graphs

Using the story, map, and graph of Jessica's travels from Lesson 6, page 62, create each of the following graphs:

- a graph showing the total distance Jessica traveled
- a graph showing how Jessica's distance from the school changed over time
- a graph showing how Jessica's distance from the park changed over time

hot words
total distance graph
distance-from graph

Homework
page 84

8 Juan and Marina Go Walking

INTERPRETING GRAPHS OF MOTION

"A picture is worth a thousand words." Have you ever heard that saying? As you will see in this lesson, graphs are pictures that can contain a lot of information in a compact form.

Write a Story to Match Four Graphs

What story can you write based on four different graphs?

Write an imaginative story that matches the information on the graphs below. Use information from all four of the graphs.

Juan's and Marina's Walks

Juan's Graphs

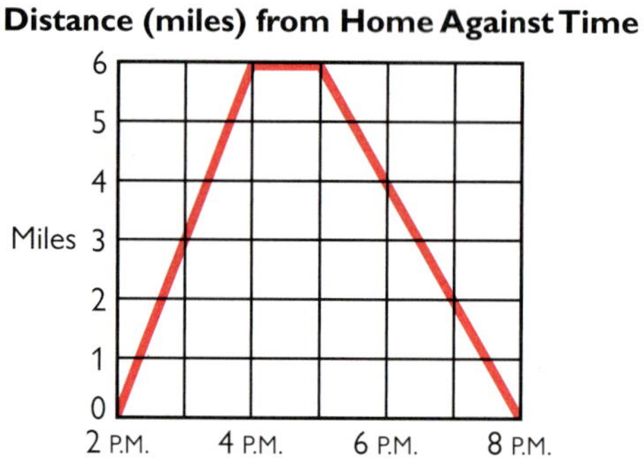

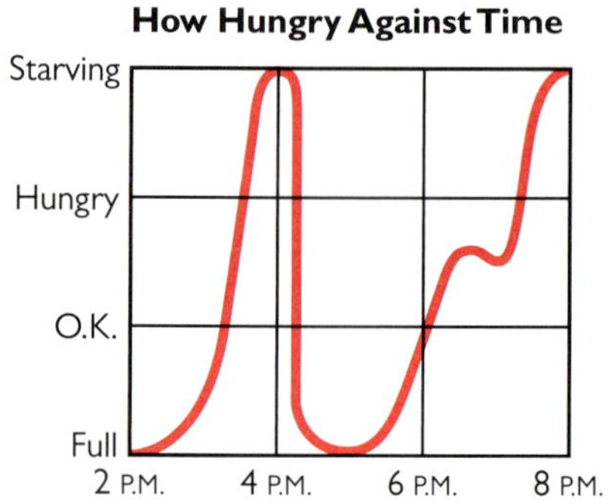

Marina's Graphs

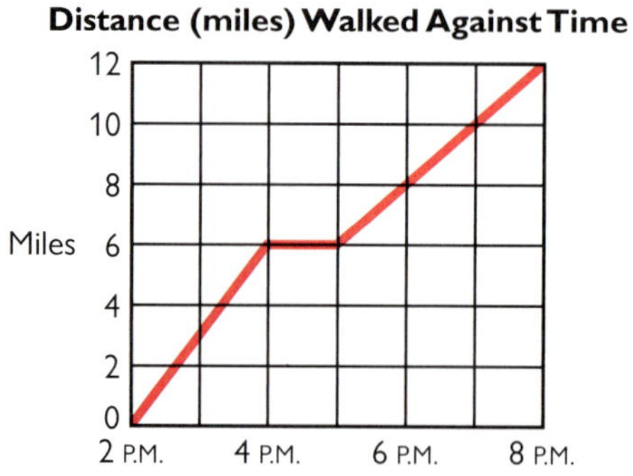

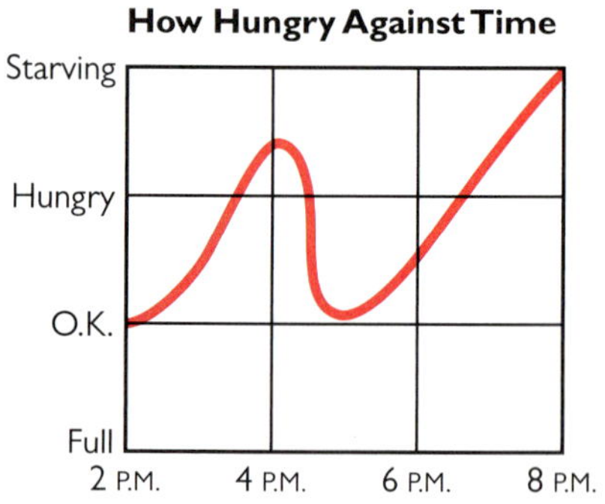

Find the Impossible Graphs

Examine the distance-time graphs shown. Then answer the following questions.

1 Which of the graphs below could show total distance traveled? How did you decide?

2 Some of the other graphs could show distance from a starting point, but some of them cannot possibly be correct! Which graphs cannot represent the journey of a single person or vehicle? Why?

3 For each of the graphs that is possible, describe in one sentence what the graph might show. For example, could any of the graphs represent throwing a ball? riding a bike?

Write Science Fiction from Impossible Graphs

Extra challenge: Make up a science fiction story that makes one of the impossible graphs possible.

Distance-Time Graphs

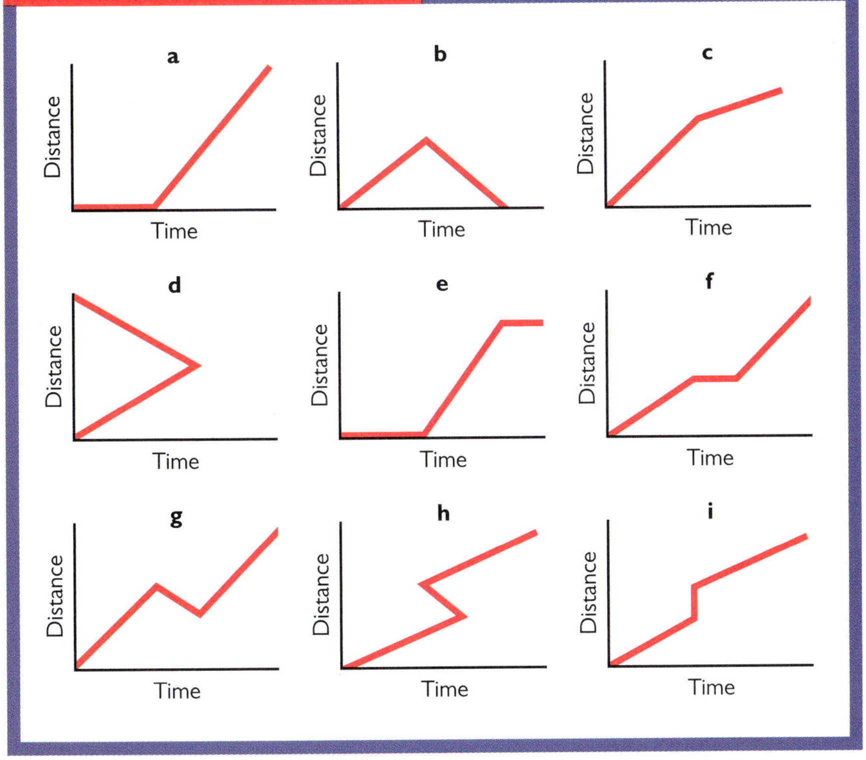

Which of the graphs shown are impossible?

hot **words** | dependent events

Homework
page 85

PHASE THREE

To: Channel KMTH Sports Announcers

The station has purchased a new graph-making computer to help you with your sports announcements. The computer creates graphs that show distance versus time and speed versus time for running races, swimming races, and other sports events. The graphs allow you to give more precise information to your television viewers. Practice reading the graphs of the races, for they will provide much interesting information—if your graph-reading skills are up to it!

Who will make it first to the finish line? Graphs and equations allow you to see patterns that help you predict the results of a race. Using mathematics allows you to do much more than predict the outcome of a race. With the techniques you will learn in this phase, you can calculate how long it will take to ride your bike someplace you have never been before, or how fast you should walk to get to your friend's house in 10 minutes.

Using Graphs to Solve Problems

WHAT'S THE MATH?

Investigations in this section focus on:

MEASUREMENT and GEOMETRY

- Determining or estimating average speeds

NUMBER and COMPUTATION

- Interpreting and creating distance-time graphs
- Interpreting and creating speed-time graphs
- Relating distance-time and speed-time graphs
- Comparing constant and variable speeds on distance-time graphs

ALGEBRA FUNCTIONS

- Applying graphs to solving problems
- Applying equations relating distance, time, and speed to solving problems
- Applying average speed to solving problems

MathScape Online
mathscape3.com/self_check_quiz

9 The Race Announcer

CONVERTING WORDS TO GRAPHS AND BACK AGAIN

In a race, do all the runners move at a constant speed? As sports announcers know, racers normally speed up and slow down as a race progresses. Distance-time graphs also show the changing speeds in a race.

How does speed vary during a race?

Play the Role of a Sports Announcer

You will receive a graph of a running, swimming, or three-legged race. Your job is to write a commentary describing the race. Then your partner will use your commentary to recreate the original distance-time graph of the race.

1 Imagine that you are the TV announcer for the race. Describe what is happening as fully and as accurately as you can. Be sure your commentary matches the graph. Include enough detail in your commentary so that your partner will be able to make a graph from it.

2 Work with a partner who wrote a commentary on a different race. Use your partner's commentary to make a distance-time graph. Try to make your graph just like the original one. When you are finished, write your answers to the following questions.

   a. How closely does your graph match the original graph? Summarize the features that your graph shares with the original graph. Then, summarize the features that are different in your graph and the original graph.

   b. Compare the original graph of the race to the commentary you used to make your graph. Describe any errors the commentary contains. How could they be corrected?

Find Out How Fast a Robot Should Run

At what speed should you program a robot so it ties Antonio in a race?

The following graph shows how Antonio bicycled in a five-mile race. Use the graph to answer the questions below.

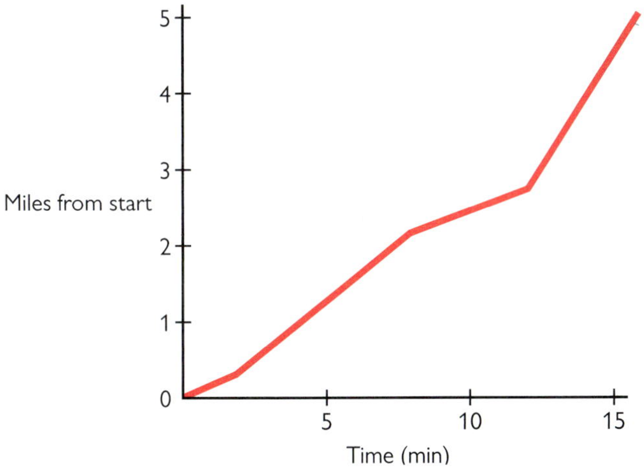

1 For which part of the race did Antonio bicycle the fastest? the slowest?

2 A robot can be programmed to roll at exactly the same speed for the entire race. How fast should the robot go if you want it to tie Antonio exactly in the race?

3 How did you figure out the speed at which to program the robot so it tied Antonio?

4 Draw a graph of the robot's race. How does it compare to the graph of Antonio's race?

5 How should the robot be programmed so that it will tie the winner of the race for which you wrote a commentary? Add the robot's race to your graph.

hot words
average
average speed
total distance graph

Homework
page 86

10 How Fast? How Far? How Long?

EXPLORING SPEED-TIME GRAPHS

Study the graphs in this lesson carefully. They contain information that you have not seen in a graph before. Your job in the activities ahead is to learn how to create and interpret these new kinds of graphs.

Problem Solve with New Kinds of Graphs

What does each graph show?

Read the story about the Lin's family trip. Then examine the four graphs of their trip and answer the following questions.

1. What does each graph show?
2. How should the vertical axis be labeled?
3. What scale should go on the vertical axis?

The Lins Go on an Outing

On Sunday, the Lin family set out on a car ride at 10 A.M. For the first hour, they drove at a speed of 40 miles per hour. In the second hour, traffic was heavy, so they only drove at 20 miles per hour. From 12 P.M. to 1 P.M., they stopped for lunch and did not drive at all. After lunch, it started to rain, so they decided to go home. They drove at 30 miles per hour to get home.

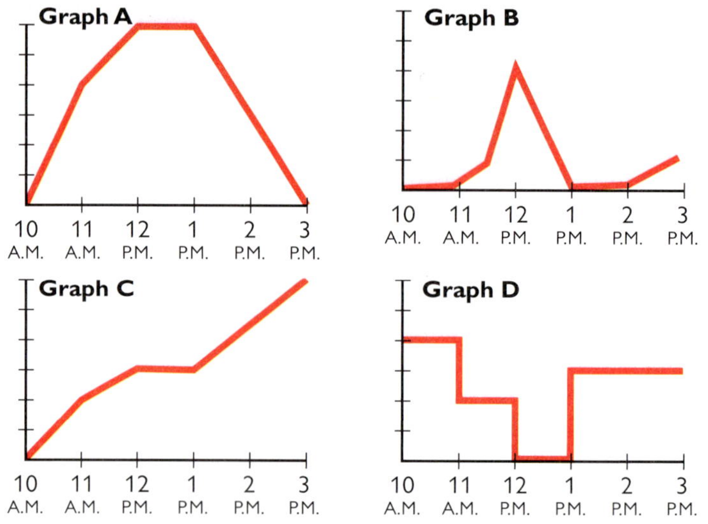

Create Distance-Time and Speed-Time Graphs

What do the graphs of some familiar motions look like?

1 Choose three different motions you have done or seen. Here are some motions you could use, or you can think up your own.

- **a.** someone going down a water slide
- **b.** someone on a bicycle climbing to the top of a hill and then coasting down the other side
- **c.** the movement of a child in a swing
- **d.** an airplane taking off
- **e.** someone who jumps out of an airplane and opens her parachute when halfway down
- **f.** an athlete in the high-jump event at the Olympics
- **g.** a baseball being pitched and then hit for a home run
- **h.** a cat trying to catch a mouse

2 For each motion, create one type of distance-time graph and a speed-time graph.

3 For each motion, also find or estimate the average speed and the fastest speed reached during the motion. When was the person, animal, or object moving the fastest?

Decide Which Graph to Use

Consider the three types of graphs about motions. For each question, say which graph or graphs you would use to find the information you need, and how you would find it. Questions 3, 4, and 5 are for graphs that show more than one person moving.

Information from Graphs of Motion

1. How far did someone travel?
2. How much time did someone spend traveling?
3. Who is traveling the fastest?
4. Is one person moving twice as fast as another?
5. When are two people moving in opposite directions?
6. What is someone's average speed over a whole trip?

hot words
slope
speed-time graph

Homework
page 87

11 The Race Is On!

SOLVING DISTANCE, SPEED, AND TIME PROBLEMS

Most things don't move with constant speed. Take your bike, for example. You start off slowly, pedal to a coasting speed, slow down soon and pedal again. Constant speed is rare. So in problem solving, you often use average speed.

What equations relate distance, speed, and time?

Problem Solve with Average Speed

Use the table to answer the following questions.

1. For each distance, what is the average speed of the world champion runner?
2. How does the average speed change as the distance of the race increases?
3. What average speed would you predict as the world record for the following women's races:

   a. 1,000 meter b. 5,000 meter c. 25,000 meter

World's Running Records for Women

Distance	Time	Runner and Year
100 meters	10.5 seconds	Florence Griffith-Joyner, 1988
200 meters	21.3 seconds	Florence Griffith-Joyner, 1988
400 meters	47.6 seconds	Marita Koch, 1985
800 meters	1 minute, 53 seconds	J. Kratochvilova, 1983
1,500 meters	3 minutes, 51 seconds	Qu Yunxia, 1993
3,000 meters	8 minutes, 6 seconds	Wang Junxia, 1993
10,000 meters	29 minutes, 32 seconds	Wang Junxia, 1993
42,206 meters (marathon)	2 hours, 15 minutes	Paula Radcliffe, 2003

Invent the Story of a Graph

The graph shows three people, animals, or objects (A, B, and C) that move over time. It shows their distances from a certain point.

1 Write a story that goes with the graph. The story can be about anything you choose—people, animals, vehicles, or asteroids. Your story should include details about the speeds and directions of the three people, animals, or objects. Also say where the distances are measured from (zero on your graph).

2 Decide on scales for the time and distance axes that match your story. Mark your scales on the graph.

3 Draw a graph that shows the total distance traveled over the given time by A, B, and C. How far did each of A, B, and C travel?

4 Draw a graph that shows the speeds at which each of A, B, and C traveled during your story.

5 Find the average speed for each of A, B, and C.

A Graph to Finish

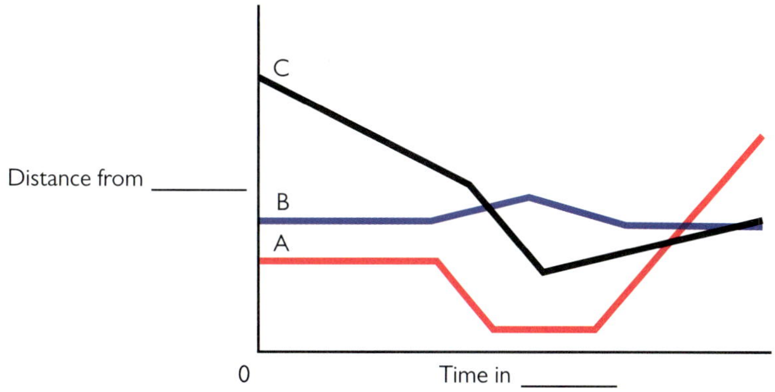

What story will the graph tell when finished?

hot **words** average speed
distance

Homework
page 88

12 Final Project

COMBINING MOTION DATA, GRAPHS, AND EQUATIONS

Here is your chance to use everything you have learned about motion to analyze, compare, and describe motions of your own choosing.

Collect the Motion Data

What kind of motion should you study?

Gather data about a motion of your own choosing.

1. Pick a situation from Sample Motions that allows you to compare the motion of two or more people, animals, or objects.
2. Collect as much data on the motion as you can. For each of the people or objects you will be comparing, be sure to find as many times, speeds, and distances as possible.

Sample Motions

- An express train and a local train that pass through your town
- The winner of the Indy 500 and the driver who came in second
- The results of two participants in your school's 100-meter race
- Olympic track and field events, such as the 100-meter sprint, the hurdles, the long jump, or the high jump
- Walking and rollerblading
- Flying from Atlanta to Los Angeles on an airplane versus taking a train
- A jaguar and a cheetah
- A jaguar and a tortoise
- The speed of a hockey puck versus the speed of a baseball
- Balls rolling down ramps with different slopes
- Bicycling in different gears

Analyze the Motion

For each of the people or objects that you are comparing, analyze the data you collected and determine each of the following.

- the total distance traveled and the total time
- the average speed over the entire trip
- intermediate speeds, whenever possible
- equations that can be used to calculate time, speed, and/or distance for each person/object
- graphs displaying total distance traveled, distance from a starting point, and speed versus time
- graphs that compare the motion of the two people/objects

Describe and Display the Motion

Prepare a chart, booklet, or other presentation that describes and displays your findings as fully as possible. Be sure to include the following:

- your initial data about the motion
- all of the equations you developed
- graphs of all motions
- maps, diagrams, or illustrations describing each motion
- written descriptions or stories for each motion

Your final project should show how much you know about motion.

hot words
total distance
(formula $d = s \times t$)
statistics

Homework
page 89

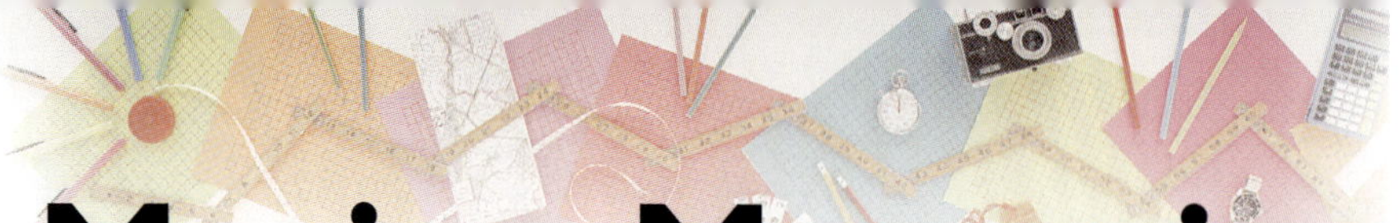

1 Homework

Moving, Measuring, and Representing

Applying Skills

For each motion, tell which of these speeds are possible: 200 meters per second, 27 feet per second, 20 miles per hour, 90 feet per second, 50 meters per hour, 500 miles per hour, 300 feet per second, 5 feet per hour.

1. a jet plane flying

2. a cheetah running at full speed

3. an athlete sprinting

4. a snail crawling

What units could you use for each motion?

5. a train traveling at full speed

6. the turning of a phonograph record

7. the rotation of the hour hand of a clock

8. a spider walking on a wall

Extending Concepts

9. Use the second hand of your watch to measure your pulse rate for 1 minute. Give the rate together with the appropriate units.

10. Find a space where you can walk for 10 seconds. Walk at your normal pace for 10 seconds and count how many steps you take. Estimate your stride length in feet. Estimate the distance that you would cover in 1 minute. Estimate your walking speed in feet per minute.

Making Connections

A *tsunami* (tidal wave) may be caused by an earthquake beneath the ocean. Its speed depends on the depth of the ocean. As it approaches shallow water along a coast, it slows and may increase in height to as much as 115 ft. Each picture below gives information about a tsunami.

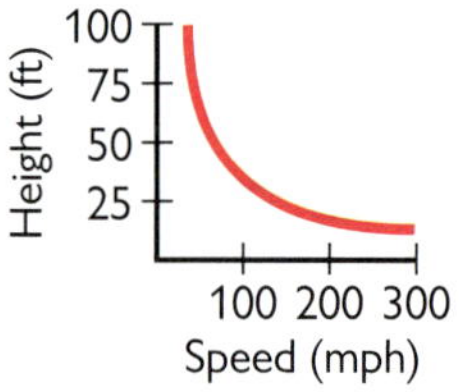

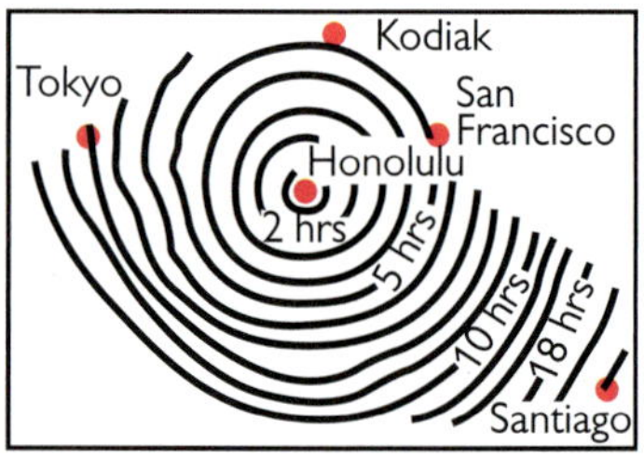

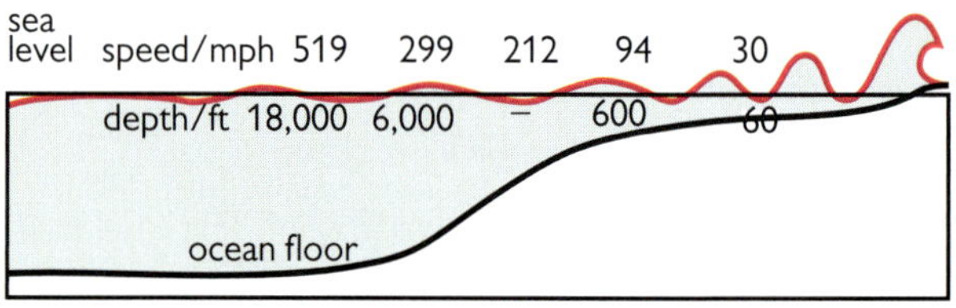

11. What information is given in each representation? For each one, give an example of a question that could be answered using that representation.

12. Which representation, if any, would you use to answer the question, "How fast was the tsunami traveling when it was 500 miles from Hawaii?" Explain.

13. Write three questions about tsunamis that cannot be answered using any of the representations above.

2 Homework

Going the Distance

Applying Skills

Estimate how far you would go in 20 seconds by:

1. walking quickly
2. hopping
3. walking backward
4. Copy and complete the table below.

	Walking	Jogging	Cycling
Distance in 10 seconds	50 feet	85 feet	280 feet
Distance in 1 second			
Distance in 1 minute			
Distance in 1 hour			
Time to go 1 mile			

5. Copy and complete the table below. Include the appropriate units.

Distance	Time	Speed
100 miles		40 miles per hour
	8 seconds	6 feet per second
160 meters	4 minutes	

Extending Concepts

6. a. Bill runs as fast as he can for 12 seconds and covers a distance of 280 feet. At that pace, how far would he run in one minute? in one hour? What is his speed in miles per hour?

b. In item **6a,** when you calculated how far Bill would run in one hour, what assumption did you make? Do you think he would really run that far in an hour? Why or why not? How far do you think he would run?

7. Daniel runs the length of his backyard in 10 seconds.

a. If Scott runs the length of the yard at half Daniel's speed, how long will it take him? How long will it take him at twice Daniel's speed?

b. If Kemi runs three times the length of the yard in 30 seconds, how does her speed compare to Daniel's?

c. If Doug runs twice the length of the yard in 5 seconds, how does his speed compare to Daniel's speed?

Making Connections

8. a. The speed of light in a vacuum is about 186,000 miles per second. How far would light travel in a vacuum in one hour? What is its speed in miles per hour?

b. The speed of a glacier depends on its steepness and its temperature. A steep, warm glacier may flow as much as 100 feet per day. At this rate, how far would it flow in a year? in one hour? What would its speed be in miles per hour?

Reporting Live from the Parade

Applying Skills

The field notes below give information about the position of characters in a parade. Use this and your Snapshot Sequence Sheets to answer the questions. The distance between each pair of landmarks on your Snapshot Sequence Sheet is 15 meters.

A = Acrobat
AD = African Drummers
CT = Capoeira Troupe

1st Snapshot, 10:30:00 A.M.
A: At Gas Station
AD: 10 meters past Gas Station
CT: At Taco Zone

2nd Snapshot, 10:30:10 A.M.
A: 5 meters past Dog Lovers' Club
AD: 5 meters past Dog Lovers' Club
CT: At Taco Zone

3rd Snapshot, 10:30:20 A.M.
A: 10 meters past Taco Zone
AD: At Taco Zone
CT: At Taco Zone

4th Snapshot, 10:30:30 A.M.
A: At *U.F.O. Chronicle*
AD: 10 meters past Taco Zone
CT: At Taco Zone

1. Use the field notes to complete the first four pictures of your Snapshot Sequence Sheet. In each picture, record the positions of the characters.
2. How far is it from the Taco Zone to the T-Shirt Hut? from the TreeWatch booth to the stop sign at Avenue B?
3. How much time is there between consecutive snapshots?
4. How much time is represented by Snapshots 1–4 all together?
5. How far does each character travel between snapshots?
6. How fast, in meters per second, is each character moving? Who is traveling the fastest?

Extending Concepts

7. Write a rule you can use to calculate how far in meters the acrobat would move in any given number of minutes. Use D to stand for the distance and M for the number of minutes. Use your rule to figure out how far the acrobat would move in 9 minutes.
8. Write a rule you can use to calculate how many snapshots would be taken in any given number of minutes. Use P to stand for the number of snapshots and M for the number of minutes. How many snapshots would be taken in 9 minutes?
9. Write your own snapshot sequence for Snapshots 5–8 to show how the parade characters might continue. Use your sequence to complete Pictures 5–8 of your Snapshot Sequence Sheet.

Writing

10. Write a short narrative describing what the three parade characters might have been doing in Snapshots 1–8. Explain how you can tell.

4 Homework

The Parade Continues

Applying Skills

The field notes below give information about the position of characters in a parade. Use this and your Snapshot Sequence Sheet to answer the questions. The distance between each pair of landmarks on your Snapshot Sequence Sheet is 15 meters.

SW = Stilt Walker, **S** = Saxophonist, **D** = Danny

1st Snapshot, 2:20:00 P.M.
SW: At T-Shirt Hut
S: At TreeWatch
D: 5 meters past Gas Station

2nd Snapshot, 2:20:05 P.M.
SW: At *U.F.O. Chronicle*
S: 5 meters past TreeWatch
D: At Dog Lovers' Club

3rd and 4th Snapshots
Too cloudy

5th Snapshot, 2:20:20 P.M.
SW: At Dog Lovers' Club
S: 5 meters past *U.F.O. Chronicle*
D: At TreeWatch

1. Use the field notes to complete Pictures 1, 2, and 5 of your Snapshot Sequence Sheet. In each picture, record the positions of the characters.

2. Figure out what the 3rd and 4th pictures would have looked like and complete Pictures 3 and 4 of your Snapshot Sequence Sheet.

3. Find the speed and direction of each character.

4. How far would each character travel in 5 minutes?

Extending Concepts

5. What happened between 2:20:10 and 2:20:15? Create a snapshot sequence with a picture every 1 second between 2:20:10 and 2:20:15. At what point did Danny and the stilt walker pass?

6. Explain how you can use a snapshot sequence to figure out:

a. the speed at which an object is traveling

b. when one object is traveling faster than another

c. what probably happened between two snapshots

Writing

7. Write a story explaining the following snapshot sequence. Explain who the characters are and what is at each marker along the road. Give the distance between markers and time between snapshots. Give the speeds and directions of the characters. Explain why they are moving that way.

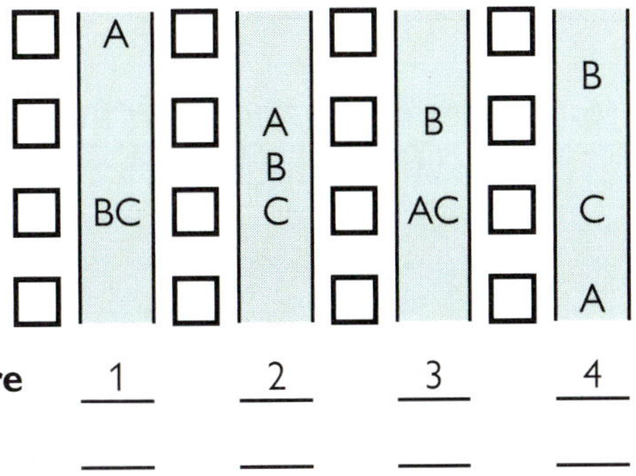

Homework 5: Walk This Way

Applying Skills

Mary walked 200 feet along a straight line in 40 seconds. Follow the guidelines to make a distance-time graph.

1. Draw axes for a graph. Label the horizontal axis "Time (seconds)" and the vertical axis "Distance from starting point (feet)."
2. Set the scale for each axis. Mark equal divisions along each axis.
3. How long did it take her to walk the entire distance? Mark that point.
4. Assuming that Mary's speed was constant, how long did it take her to walk half the distance? Mark that point.
5. Draw the line showing the entire walk.

Extending Concepts

The graphs for items 6–9 all represent motion along a straight line.

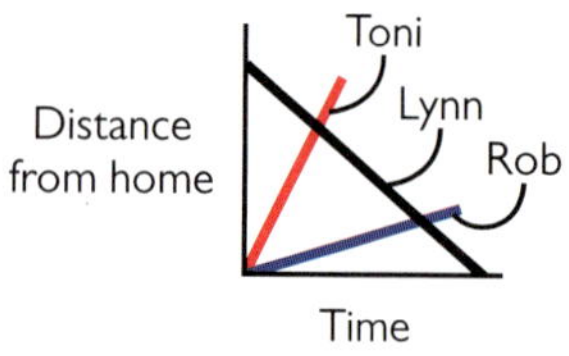

6. In Toni, Lynn, and Rob's graph, who walked the fastest? the slowest? How can you tell?
7. Why does Lynn's graph slope downward?

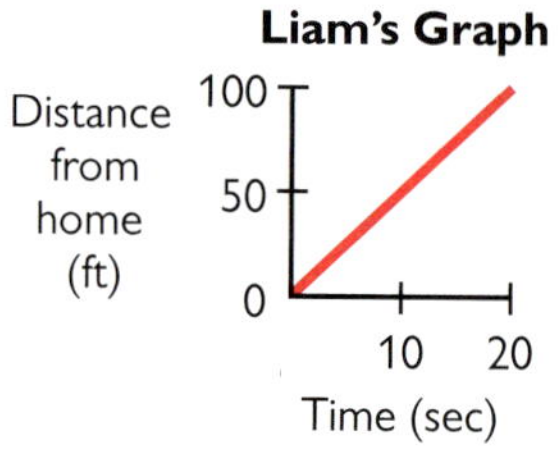

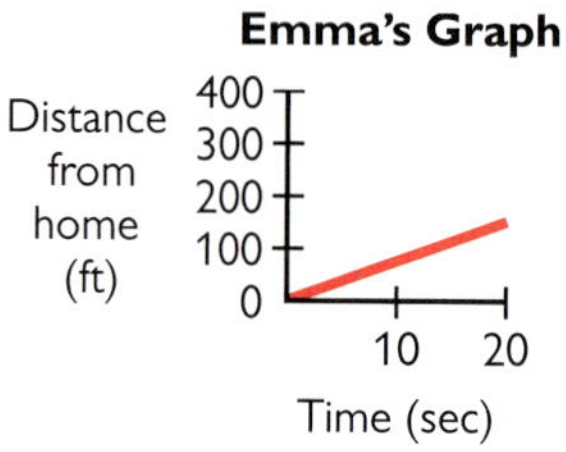

8. How far did Liam and Emma walk in 20 seconds? Who walked faster? Whose graph looks steeper? Why is it not true in this case that a steeper graph indicates a faster motion?
9. Which graph, Liam's or Emma's, has a more suitable scale? Why?

Tyler set out from the trailhead at 9:00 A.M. He walked for 1 hour, going 3 miles, and then rested for 20 minutes. He continued a little more slowly and reached Mirror Lake at 12:20 P.M. Suzi left the trailhead at 10:00 A.M. She walked steadily without breaks and also reached the lake at 12:20 P.M.

10. Create a graph to show what happens in the above story. Be sure to show the distances and times for each person. Estimate how far each person walked. Assume that each person walks along a straight line.

Making Connections

11. The Pronghorn antelope is the swiftest North American mammal. It can run 200 meters in about 7.5 seconds. The fastest human can run 200 meters in about 20 seconds. Using the same axes, make distance-time graphs for a Pronghorn antelope and a human.

Stories, Maps, and Graphs

Applying Skills

Use the story, map, and graph below to answer the questions. Tell whether each answer is exact or approximate and whether you used the story and map below or graph in column 2 to find it.

It was Friday evening. David remembered that his sister's birthday was the next day. He decided to buy her a CD. He left his house on Hill Street at 5:30 P.M. to walk to the store. A few moments later, he ran into a friend and stopped to chat for a while. When he left his friend, he remembered that the CD store would close at 6:00 P.M., and he ran the rest of the way. He reached the store in time, bought a CD, and walked home.

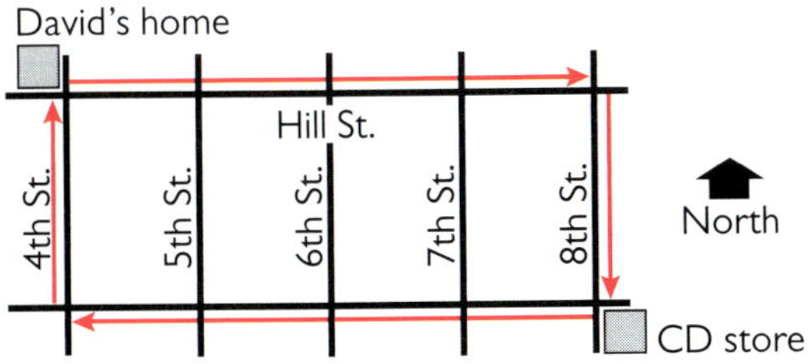

1. At what time did David return home?
2. How many blocks did David walk along Hill Street before turning?
3. Why did David want to buy a CD?
4. When was David traveling the fastest? the slowest?
5. Why did David hurry as he left his friend?
6. How long did David spend talking to his friend?
7. In what direction was David traveling when he first left his house?
8. During what time period(s) was David moving closer to his house?

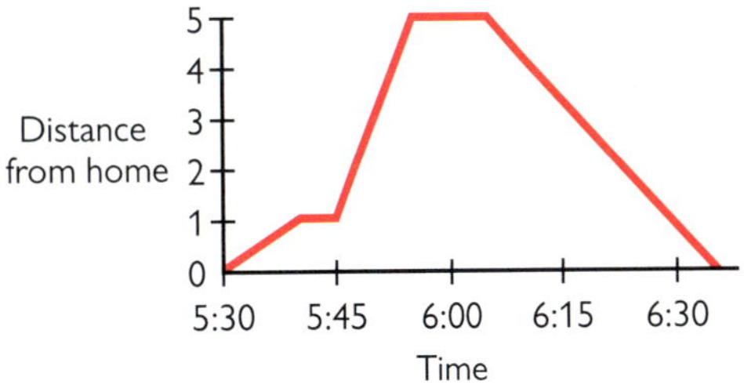

Extending Concepts

9. Write a story to describe a fantasy trip. Make a map and a distance-time graph to go with your story. Then make a second graph showing how a character in your story felt during the trip.

Writing

10. Make up a question about your story that can be answered most easily using your map. Make up another question that can be answered most easily using your distance-time graph.

A Graphing Matter

Homework 7

Applying Skills

John walked 100 feet in a straight line from the starting point to the ending point in 15 seconds, then 40 feet back toward the starting point in 5 seconds. He waited 5 seconds, then walked all the way back to the start in 10 seconds. His "Distance from Start" graph is shown here.

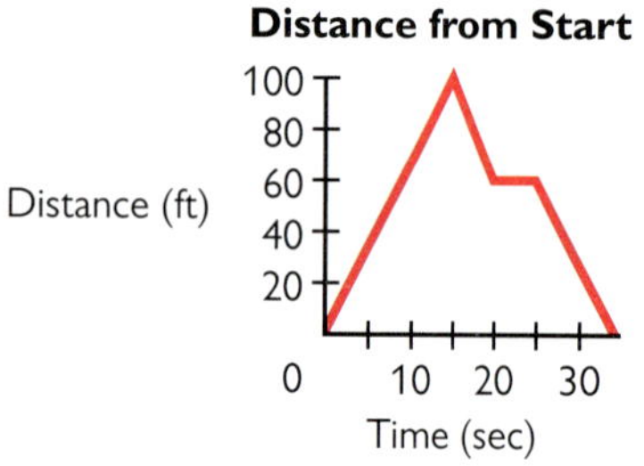

Follow the guidelines to make John's "Distance to End Point" graph.

1. Draw axes for a graph. Label each axis appropriately. Set the scale.
2. Find John's distance to the ending point after 0, 15, 20, 25, and 35 seconds. Mark those points. Draw the line showing the entire walk.

Follow the guidelines to make John's "Total Distance" graph.

3. Draw axes for a graph. Label each axis appropriately. What is the total distance that John walked? What will be the largest number on the vertical axis? Set the scale on each axis.
4. Find the total distance John had walked after 0, 15, 20, 25, and 35 seconds. Mark those points. Draw the line showing the entire walk.

Extending Concepts

5. The three graphs at right show total distance, distance to ending point, and distance from starting point. Which is which? How can you tell?
6. Which two graphs could represent the same motion? For each pair of graphs that could *not* represent the same motion, explain why not.

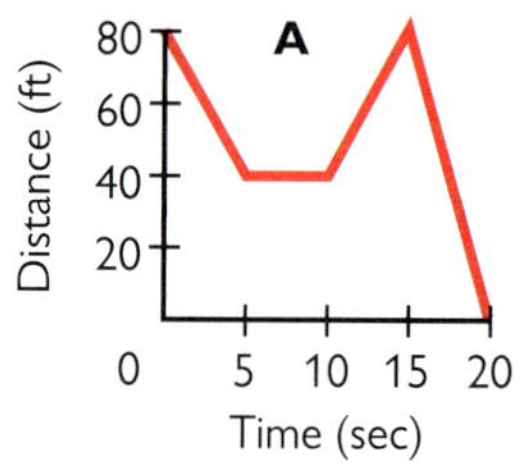

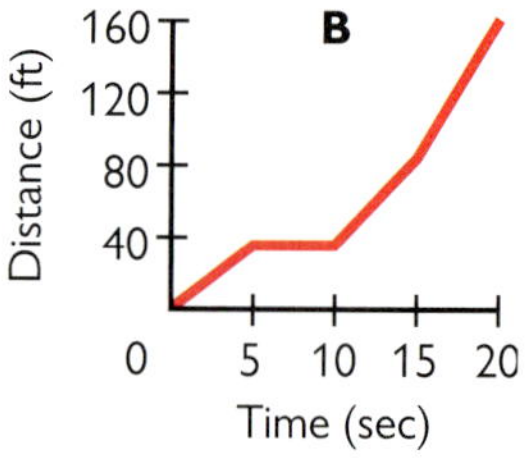

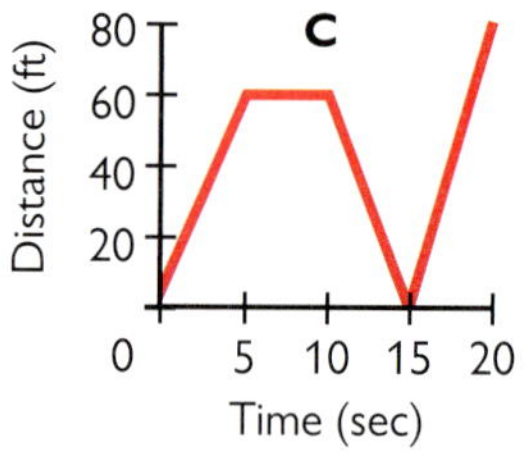

Making Connections

The gray whale migrates seasonally between its breeding ground in Baja California and its feeding ground in the Arctic. The graph shows distance to the Arctic for the gray whale for a one-year period.

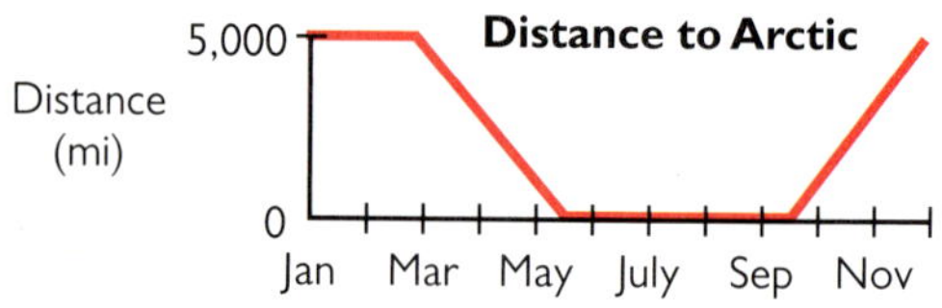

7. A part of the graph is sloped upward. What are the whales doing during this time period? Where are they during February? How can you tell?
8. Draw the corresponding "Distance from Baja" and "Total Distance" graphs. Explain how you figured them out.

Homework 8

Juan and Marina Go Walking

Applying Skills

a.

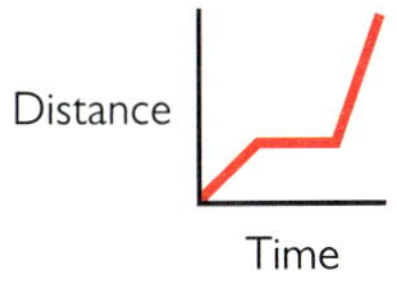

b.

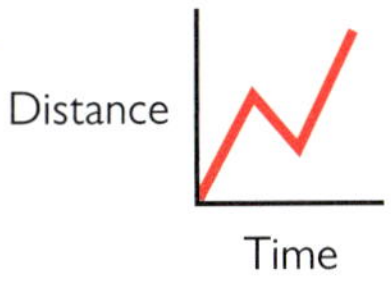

c.

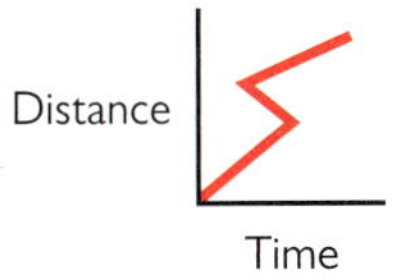

d.

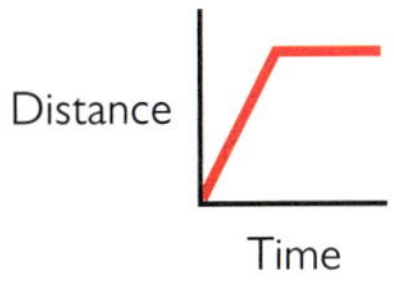

1. Which of the graphs above could show total distance traveled? Why?

2. Which graphs could not show the journey of a single person or vehicle?

3. Which graphs could represent distance from a point?

4. Sketch a distance-time graph that could show total distance traveled.

Extending Concepts

Leah's Graph
Distance (mi) from Home Against Time

Jamie's Graph
Distance (mi) Walked Against Time

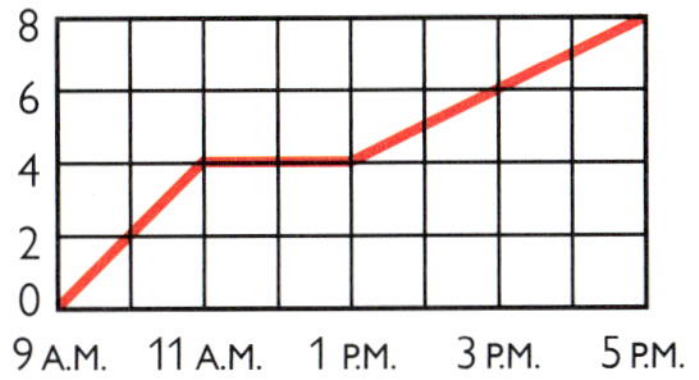

5. What is graphed on the horizontal axis of each graph shown? on the vertical axis? Are the scales of the two graphs the same?

6. How far did Jamie walk from 1 to 5 P.M.?

7. Why is there a flat portion on Jamie's graph between 11 A.M. and 1 P.M.?

8. Why does Leah's graph slope up and then down?

9. Can you tell where Jamie was at 5 P.M.? Why? Where was Leah at 2 P.M.?

Explain why each story is inconsistent with the graphs in Extending Concepts.

10. "Leah and Jamie left home at 9 A.M., walked together until 11 A.M., rested for two hours, and walked home together."

11. "Leah and Jamie left home at 9 A.M. Jamie walked faster than Leah until 11 A.M. Both rested between 11 A.M. and 1 P.M. Then they walked home."

12. "Leah and Jamie left home at 9 A.M. and walked 4 miles north. After lunch they walked another mile north. Then Jamie walked to his friend's house, and Leah walked home."

Writing

13. Write your own imaginative story that matches the information on the two graphs in item **5** and the graph below. Use information from all three graphs.

Jamie's Graph
How Tired Against Time

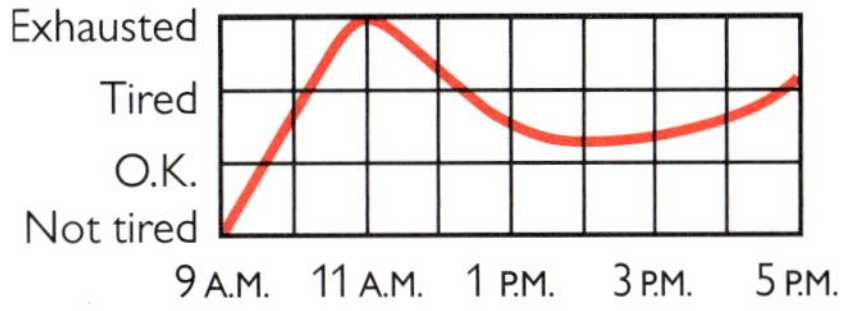

The Race Announcer

Applying Skills

Refer to the graph of the running race for items 1–4.

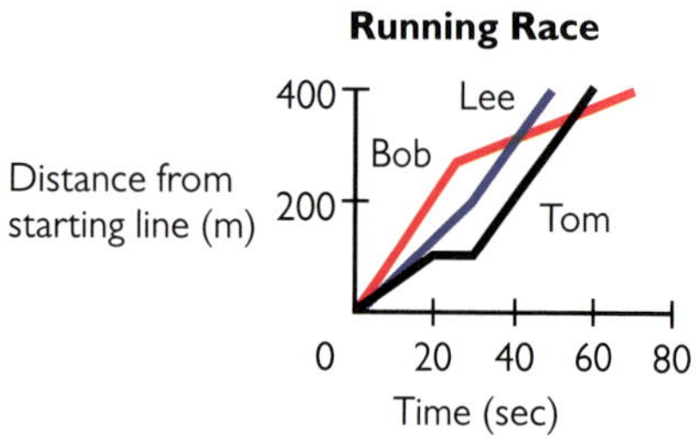

1. At what times were there changes in speed?

2. At what time did Lee pass Bob?

3. What were the relative positions of the racers after 10 seconds?

4. How long did each person take to complete the race? Who won?

Refer to the graph of the swimming race for items 5 and 6.

5. Who is in the lead at 20 seconds? at 40 seconds? at 58 seconds? What is happening at 30 seconds?

6. How long does each person take to complete the first 30-meter lap? Who completes the first lap more quickly?

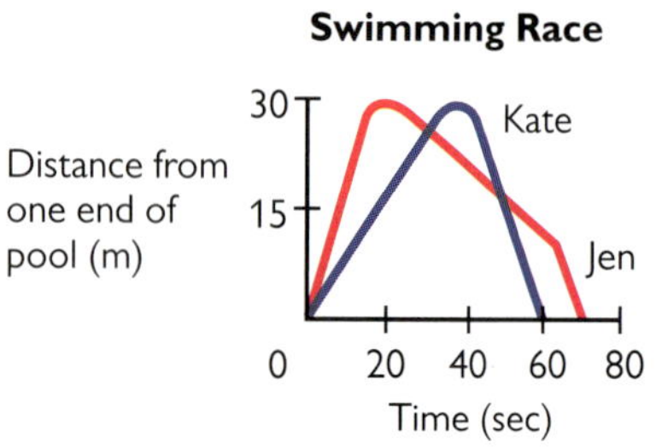

Extending Concepts

7. a. In the running race, what is the total distance that Tom runs? What is his total time? his average speed?

b. What is Tom's speed at 10 seconds? 25 seconds? 40 seconds? Is there some point in the race when he was running at the average speed that you calculated in item **7a**? Explain.

c. If you program a robot to roll at a constant speed in the race, at what speed should it roll to tie Tom? Draw a graph of the robot's race.

d. Find Kate's average speed in the swimming race. How did you find it?

The report below doesn't give enough information to tell the final outcome. Draw two graphs that match the report but show different outcomes. Rewrite the report so that it fully and accurately describes one of your graphs.

8. "As they start on this 200-meter race, Sam is in the lead, Eric is behind him, then Raphael. Now, just 10 seconds into the race, Eric has tripped, and he's down. Raphael has passed him. This is amazing, with 50 meters to go Raphael passes Sam to win the race in 25 seconds."

Writing

9. Use the graph in item **5** to write a commentary for the swimming race. Describe what is happening as fully and as accurately as you can.

10 Homework

How Fast? How Far? How Long?

Applying Skills

For each question, tell which graph or graphs shown here could be used to find the information. Then answer the question.

1. How far was each person from home at 3:00 P.M.?
2. Who was traveling faster in each time period?
3. What was each person's average speed for the whole trip?
4. What was each person's exact speed at 4:30 P.M.?
5. During which time period(s) was Brad moving twice as fast as Amy?
6. How far did each person travel?

Distance-Time Graph

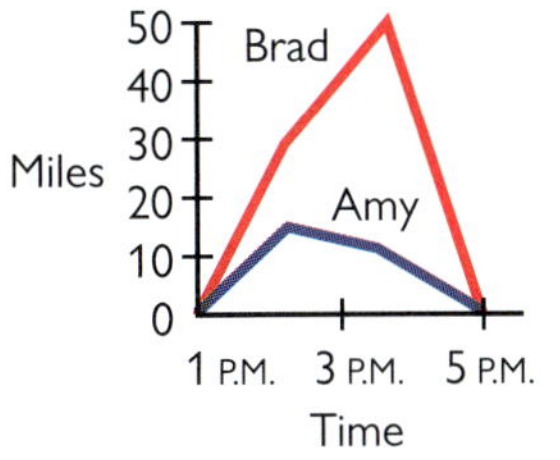

Total Distance Graph

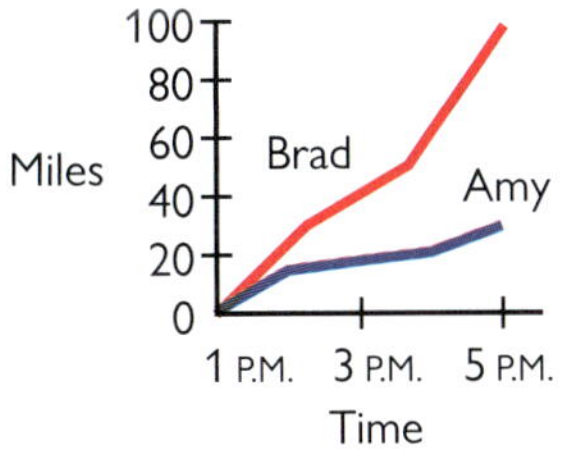

Speed-Time Graph

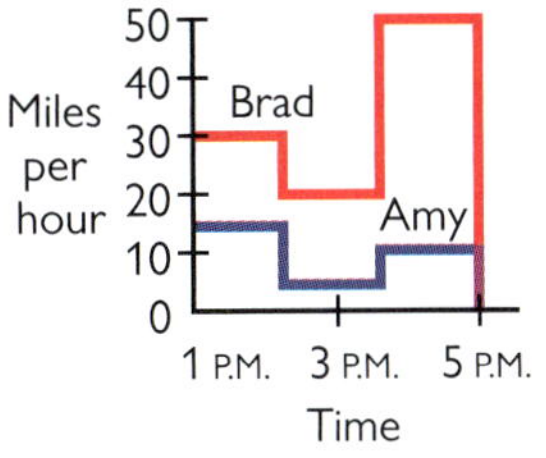

Extending Concepts

7. Find Thomas's speed during each time period. How did you figure it out? Make a speed-time graph to represent Thomas's trip.

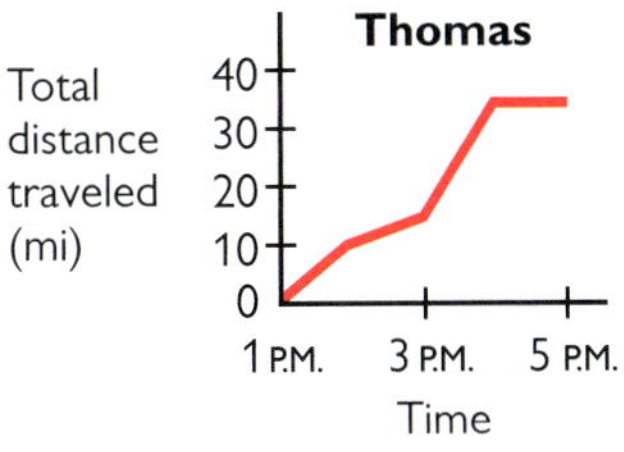

8. If a speed-time graph stays flat, what does that tell you about the motion? What will the corresponding distance-time graph look like? If a speed-time graph slopes upward, what does that tell you about the motion? Why are there no sloping lines on your speed-time graph in item 7? Do you think this is realistic? Why or why not?

Making Connections

9. An object dropped off a building accelerates because of gravity. Each second its speed increases by 32 feet per second—after 1 second its speed is 32 feet per second, after 2 seconds it is 64 feet per second, and so on. Which graph below could be a speed-time graph for an object dropped off a building? Which could be a distance-time graph? How you can tell?

a.

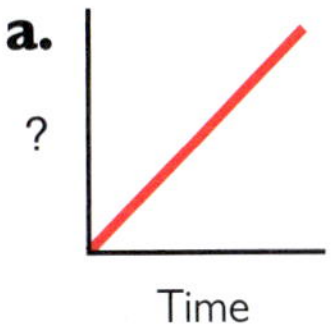

b.

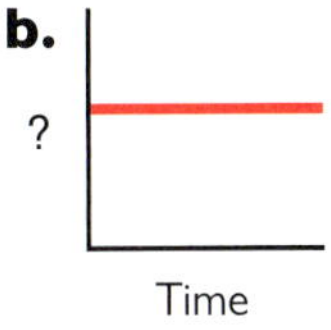

c.

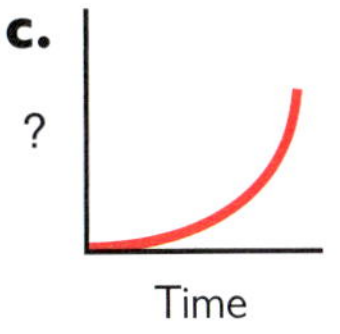

d.

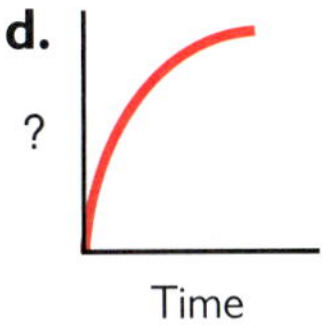

11 Homework

The Race Is On!

Applying Skills

Below are the world's running records for men in different distances.

Distance	Time
100 m	9.8 sec
400 m	43.2 sec
800 m	1 min, 41 sec
1,500 m	3 min, 26 sec

1. For each distance, find the average speed of the world champion runner.

2. How does the average speed change as the length of the race increases?

3. What would you predict for the world record for the men's 600-meter race? 1,200-meter race? 4,000-meter race?

4. Joel drove for 30 minutes at 40 miles per hour, for 1 hour at 60 miles per hour, and for 20 minutes at 54 miles per hour.

a. How far did he travel at each speed?

b. What was his total distance? his total time? his average speed?

c. Create a total distance graph for Joel's trip.

Extending Concepts

5. In a 20-mile bicycle race, Lois gave Sarah a head start of 15 minutes. Lois rode at an average speed of 22 miles per hour. Sarah rode at an average speed of 18 miles per hour. Who won the race? How did you find the answer? Make a total distance graph to show the race.

Writing

6. Write a story to go with the speed-time graph shown here. Give speeds, distances, and directions for each person. Mark scales on the graph to match your story.

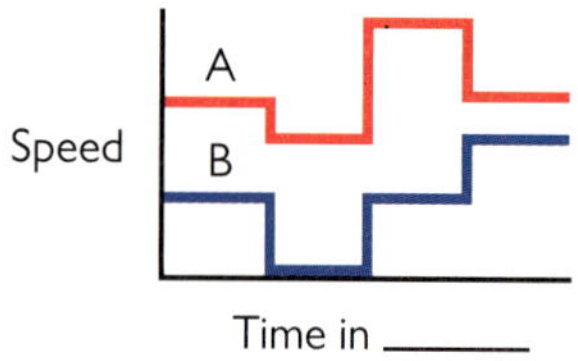

Draw a graph showing how the distance of each person from a certain point changes during the story. Tell where the distances are measured from. Find the average speed for each person.

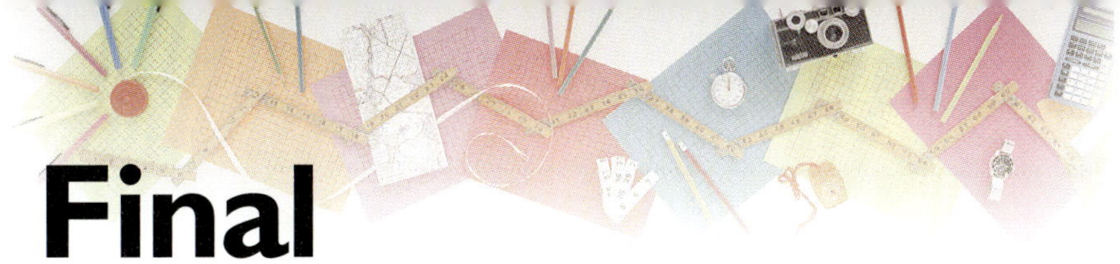

Homework 12

Final Project

Applying Skills

Felix hiked up a mountain. During the first 3 hours, the trail was steep and he hiked only 4 miles. To hike the remaining 2 miles to the top took him 1 hour. It took him 2 hours to hike down again.

1. Find the total distance he walked and the total time.

2. Find his average speed for the whole trip.

3. Find his speed for each part of the trip.

4. For each part of the trip, find an equation that can be used to calculate distance traveled.

5. Make a speed-time graph, a graph showing the total distance traveled, and a graph showing distance from the starting point.

Extending Concepts

Jim and Rajan ran a 400-meter race. Jim ran at 6 meters per second for the first 200 meters, 9 meters per second for the next 100, and at 10 meters per second for the last 100 meters. Rajan ran the first 200 meters in 22 seconds, the next 100 in 18 seconds, and the last 100 meters in 22 seconds. Use this data to write a report. Include the following:

6. For each person: the time and speed for each section of the race and the total time and average speed for the whole race.

7. Graphs showing total distance traveled and speed versus time.

8. Who was going faster during each period, at what times one person was passed by the other, and who won the race.

9. Equations that can be used to calculate time and/or distance for each person.

Making Connections

Artificial satellites, which are placed in orbit around the earth, may be used for communications, research, meteorology, or navigation.

10. Suppose a satellite travels in a circular orbit about the earth at constant speed. Without showing any scale on the axes, sketch graphs showing total distance traveled, distance from the earth, and speed versus time while the satellite is in orbit.

11. Can you assume that an object is not moving if its distance-from graph is flat? if its total distance graph is flat? Explain your thinking.